P9-ELX-785

PET OWNER'S GUIDE TO
COLDWATER FISHKEEPING

Andrew Eade

ABOUT THE AUTHOR

Andrew Eade has been interested in all things aquatic since childhood, first investigating rock pools, rivers and streams near his home, and later breeding a variety of fish species with considerable success. Following academic studies in related sciences, Andrew spent a period in the aquatic retail trade. He then joined the staff of Europe's leading fishkeeping magazine, *Practical Fishkeeping*, before setting up his own Scuba-diving company.

PHOTOGRAPHY

PETER GATHERCOLE
With contributions from Dave Bevan.

Published by Ringpress Books Limited,
PO Box 8, Lydney, Gloucestershire,
GL15 4YN, United Kingdom.

First published 1999

ISBN 1 86054 072 4

Printed in Hong Kong through Printworks Int. Ltd.

CONTENTS

1 Introduction To Fishkeeping

Welcome to the wonderful world of fishkeeping. Whether you already have your first aquarium, or you are considering setting one up, you will already be aware of just how exciting, fascinating and absorbing fishkeeping can be. With this book I hope to help smooth the way from your first teetering steps before you fall headlong into fishkeeping. It can be a tortuous journey from inexperienced novice to experienced fishkeeper, but with

an open mind and a little help, problems will be few and far between.

In this book we cover every conceivable angle of coldwater fishkeeping, from choosing and setting up your tank to selecting your fish. Once you have your fish, what kind of care do they need? What if they breed? Fear not, we have covered all these questions and more. So, without further ado, roll up your sleeves and delve into fishkeeping – you will not regret it.

WHICH FISH?

As soon as you walk around your first aquatic shop you will be astonished at the variety of fish available. To the uninitiated it can be quite bewildering learning about which fish can be kept together, but the aim of this section is to break the fish down into groups and give you a little insight into the needs and requirements of each of them. In general, there are three main groups of fish – tropical, coldwater and marine. This book is concerned with the correct care and upkeep of coldwater fish, so we will take a closer look at them.

GOLDFISH

The original goldfish was a dull, brown colour, closely resembling a Crucian carp which is the wild version of the species. In ancient China these fish were bred for their food value in vast numbers in pools closely resembling rice fields. Some sixteen hundred years ago, fish farmers in China noticed that some of their carp were developing gold coloration. These fish were taken and intensively bred.

The goldfish has been selectively bred over many centuries.

Through trial and error, completely orange fish were produced and the modern goldfish was born. These great golden fish were, and still are, revered in the ancient history and mythology of China and much of South Eastern Asia.

There are examples of pottery and sculpture from nearly fifteen hundred years ago which depict the goldfish as an almost mythical creature. Strains of these wonderful creatures were first recorded as imports into Europe in the early 1600s and it is possible to think that, somewhere, there are some very distantly related fish to those first, pioneering goldfish.

The Common goldfish that we have today has not changed from those first fish. Increasing breeding technology has helped to develop the finnage and coloration of the goldfish to offer us a wide choice of colour, fin shape and body shape. It has turned into a very exacting science for those people involved, and the search for the perfect goldfish still goes on. Careful selection takes place to pick exactly the right pair of fish which display excellent colour, good size, good finnage and good deportment. The combination of all these things produces some truly magnificent fish which really have to be seen to be believed. In addition to its beauty, the goldfish still remains one of the easiest fish to keep. It is hardy, easy to feed and fairly undemanding with its water requirements.

MIX AND MATCH

If you want a goldfish a little different to the normal orange critter, you need look no further than most aquatic stores, as they often carry around ten to twenty varieties of fancy goldfish. From the humble goldfish to the elaborate Bubble eye, they are all available for the home aquarium. When buying fancy goldfish, be aware that not all of them will mix together without some kind of upset. It all boils down to the gritty determination of all fish to breed. To a standard goldfish, a nice plump fancy goldfish is the sexiest thing in existence, and a fish with which it must breed. To this end, it will pursue the fancy goldfish, male or female, to such a point that the fancy is completely stressed. There is nothing that can be done, other than avoiding keeping fancy goldfish and standard goldfish together in the same tank.

Ideal tank mates are Blackmoors, Bubble Eyes, Celestials and Ranchus. Another combination is Veiltails, Ryukin, Wakin and Orandas including Pearlscales. If you bear in mind that slim, fast-moving fish do not mix with slower, fatter ones, you should have a tank with trouble-free inmates. Scavenging fish, such as Weather Loaches, can also harass fancy goldfish and so are best avoided. Let us take a look at just some of the varieties available.

VARIETIES OF FANCY GOLDFISH

Looking at these fish it is hard to believe that they are descendants of the Crucian carp. Body and fin shape have been pushed to the limits with these fish, to produce a plethora of weird and wonderful fish. Some have massively developed sacs under their eyes, while others have highly developed heads and ornate finnage. Due to the somewhat contorted body shape, these fish can experience some problems with their digestive systems, causing the fish to float rather uncontrollably around on the surfacc. This can be avoided by careful control of the fish's diet and precise temperature control. These generally are the only problems that fancy goldfish encounter and you should certainly not be put off from keeping them. There are many varieties that you are likely to come across in the shops. Here are a few examples for you.

BLACKMOOR

The scales of the Blackmoor have a velvet-like quality and the eyes protrude from the side of the

The Blackmoor.

The Bristol shubunkin has more developed fins than the London shubunkin.

head. Well-developed examples have a plump, round shape to them and lovely full finnage. They are not tolerant of low temperatures, so the water must not be allowed to drop below 65° F. An orange or brown 'Blackmoor' is usually called a 'telescope eye'

SHUBUNKIN

Two varieties exist of the Shubunkin, the London and the Bristol. Shubunkins are basically goldfish with calico coloration which may vary from predominately blue to orange with several spots of black, brown, or white. The London shubunkin has quite short fins and appears stocky in shape. It very loosely resembles the common goldfish. In contrast, a good Bristol shubunkin has far more greatly developed and rounded fins, especially the caudal or tail fin. which flow and give the fish a 'floaty' appearance. Both fish can attain a size of some six inches and are quite happy to live in indoor tanks or the pond.

RYUKIN

This is a truly beautiful example of a fancy goldfish. The body has been greatly shortened and the back is steeply curved, giving a lovely shape to the fish.
A healthy example should have a

The Ryukin is surprisingly hardy.

straight and erect dorsal fin that stands up like a flag. The tail is triple or quadruple and flowing and, when seen from behind, should resemble a butterfly. The predominant colour for these fish is orange and white, although several varieties do exist. Despite its fragile appearance, the Ryukin is considered hardy among fancy goldfish enthusiasts.

ORANDA

These are by far the commonest form of fancy goldfish seen in the shops. Selective breeding has produced this fish with a large 'brain-like' growth – the 'cowl' or 'wen' – covering the whole of the

Orandas come in different colours – this is chocolate coloured.

head. This can take several years to develop properly but looks very impressive when fully formed. A good example of an Oranda should be nicely round in shape and not elongated, the fins should be full and flowing. Any fish you buy should look alert and active with good intense coloration. As the Oranda approaches its adult size of eight inches it can become very tame and will happily suck small pieces of food from your fingers. There are many colour varieties of the Oranda, ranging from the common red Oranda to the red and white Red Cap Oranda and even calico examples.

BUBBLE EYE

A peculiar-looking creature with large fluid-filled sacs beneath each eye. These wobble around as the fish swims, giving the fish a very comical appearance. The sacs are understandably quite delicate and sharp objects in the tank should be removed to avoid accidents. Again, many colours are available but most have slightly metallic scales.

LIONHEAD

A good specimen should resemble an egg in shape. Lionheads have the same hood as the Oranda yet lack the dorsal fin, and the fins are generally shorter than the Oranda. The extent and size of the hood may vary but, as a rule, the male's hood is slightly better developed than the female's. This should not be relied upon for accurate sexing of the fish. As with all the fancy

The Bubble eye has large fluid-filled sacs under each eye.

An immature Lionhead.

goldfish, there are several colour varieties available.

POMPOM

The Pompom has distinctive cheerleader pompoms on its nostrils. These 'nasal boquets' give it its name. The true Pompom resembles an Oranda complete with dorsal fin and fully-developed hood. Many colour varieties exist but the majority of the fish available are orange or white in colour. Often the Pompom has metallic scales which glint and sparkle under the aquarium lights.

OTHER COLDWATER SPECIES

Most of the other species of fish suitable for a coldwater aquarium are collected from Northern China or North America where the climate is very similar to the UK. These fish are able to tolerate a wide range of temperatures but will be happiest in a tank with a temperature of up to 70 degrees F (21 degrees C). Often the habitats they come from are fast-flowing mountain streams, or lakes and pools fed by mountain streams. Invariably, the water for these comes from melt water higher up in the mountains and is therefore quite cool.

Alternative coldwater fish often have to be searched out, as they are not always commonly imported. That aside, they are well worth looking out for and more than worth the effort of keeping them.

Beware of the new licensing regulations for some of these species. They must never be released into British waters.

EUROPEAN SPECIES

STICKLEBACK

This is a fish that many of us encounter as children, but have you ever considered keeping them as pets? The male is by far the most colourful, with his fine breeding livery of red and blue. They are also proficient builders and can construct very complicated nests in which they house their eggs and fry. They are truly adaptable fish and will tolerate a whole range of water conditions. They can be commonly found in brackish pools, creeks and streams and have even been seen nesting next to a seaside breakwater in full salt water. As they can get aggressive, keep them in their own tanks.

LOACH

There are a few species that are readily available to the fishkeeper which are suitable to be kept in coldwater aquaria. Loaches are generally bottom-feeding fish and are very useful for eating uneaten food after feeding time. The Weather loach is a peculiar little

A three-spined Stickleback.

The Weather Loach is very sensitive to climatic changes.

character that is extremely sensitive to changing weather conditions. This usually placid fish reacts dramatically to weather changes and will start to swim relentlessly around the tank should there be a changing weather front on its way. They are of a peaceful nature and are an ideal addition to any coldwater tank. The Striped loach (*Botia striata*) is only a good choice for a tank containing fast-moving fish as it can be a little boisterous at times. This can be alleviated by keeping more than one of them in a tank. They appreciate a good mixed diet, including tablet food and frozen food.

BITTERLING

A gorgeous little fish which has a very unusual breeding habit. It uses a Swan mussel (the 'Painter's' mussel) to incubate its eggs. The drab female fish develops a long tube, known as an ovipositor, which is inserted into the gill opening of the mussel. The eggs are laid here and hatch after four to six days. The minuscule fry are then 'spat' out of the mussel carrying even smaller mussel larvae attached to their gills. The males exhibit wonderful coloration, including hues of red and blue. A little-known benefit of keeping Bitterling is their liking for nibbling annoying black hair algae. This type of algae seems to be unpalatable to most fish but the Bitterling seems to positively relish it. They have been used to good effect to clear hair algae in both tropical and coldwater tanks.

GUDGEON

This is another 'tiddler' common in British streams. The subtle patterning of blue dots is really accentuated if the tank receives a little natural light. Its streamlined shape and sensory barbels are the perfect devices for detecting food among the gravel. This is a good fish to have in the tank as it helps to keep the tank clear of uneaten food, but needs a very clear, well oxygenated tank.

RUDD

This is a native British fish which is often offered for sale. Young fish are fast-moving with red finnage. They take most of their food from the surface. Their upturned mouths are designed for the job. They grow quickly into lovely

The streamlined Gudgeon finds food among the gravel.

The Rudd has beautiful deep red finnage.

bronzed, plump fish with deep red finnage. They are best suited to a pond, as they can be quite skittish by nature and will grow to a maximum size of twelve inches.

KOI

This is a beautiful coloured carp whose name translated from Japanese Nishikoi means 'brocaded carp'. In the right conditions these fish really grow fast, attaining at least 12 inches a year until they reach their maximum size of 36 to 40 inches. These are only suitable for a tank for the first year to eighteen months. If your tank is large enough, they can be kept indoors for longer, but carefully watch the water conditions.

PERCH

The king of freshwater British fish, the Perch has one of the most magnificent appearances of any fish swimming in British waters. Juveniles have the most defined banding, the black bands contrasting well against the green base colour and red fins. When the spiky dorsal fin is held aloft, the fish has the appearance of a formidable predator

Koi can be kept in an aquarium for the first 12-18 months.

which, indeed, is what it is. No small fish are safe with a Perch in residence. Its jaws are capable of stretching its mouth to enormous proportions. They will feed on all manner of foods including flake and bloodworm, and larger fish will enjoy frozen lancefish and earthworms. If you decide to keep these fish, set the tank up for Perch only. Introduce a small shoal and expect one or two to disappear as the smaller ones are picked off. Adults can be up to 24 inches long; a common size is around 12 inches, so a large tank is needed.

MILLER'S THUMB OR BULLHEAD

This is a tiddler which can be caught using the bottle trap method from any fast-flowing stream with good water quality. They prefer to spend much of their time under stones where they wait for food to pass by. Spawning takes place under a stone, where a number of orange eggs are laid on the underside of a stone. The male fish spends the incubation period looking after these eggs until they hatch. The fry absorb their yolk sacs and are then allowed to fend for themselves. They rarely exceed three inches in length and are interesting to keep due to their unusual breeding habits. Like the Gugeon, they need a cool, well oxygenated tank.

The prehistoric-looking Sterlet needs plenty of space, and will outgrow the average aquarium.

STERLET

A recent rise in the popularity of this fish has encouraged the importation of several species and now around six varieties are available. Most of them grow large and are better suited to life in a well-filtered pond where they can attain their true potential. They look fantastic, their prehistoric shape remaining unchanged for millions of years. They use the sensory barbels under their snout to detect small insects and crustaceans in the substrate. Although not specifically predatory, they will take small fish should they come across any. Depending on species, they can attain a length up to 72 inches, although the common species attain a size of 24 to 36 inches. As such, they are not really suitable for most coldwater aquaria.

TEMPERATE SPECIES

Some of the common tropical species can tolerate life in a coldwater tank as long as the temperature does not drop below 65 degrees F (18 degrees C). These include Guppies, Three spot gouramis, Paradise fish, Zebra danios, White Cloud Mountain minnows and Peppered catfish. They should be kept in a tank with small coldwater fish, as none of them exceed three inches in length.

PARADISE FISH

One of the first tropical/temperate fish to be imported into Europe. They are attractively coloured, displaying many shades of blue and red. They can be quite aggressive and are better kept away from fancy goldfish as they may attack the long, flowing fins. They have a very interesting breeding method whereby they build a complex nest of bubbles on the surface of the water. The eggs are painstakingly blown one by one into the nest by the male fish and guarded until they hatch. They prefer to be kept in a tank with quite substantial plant growth where they can look truly stunning swimming against a natural backdrop of plants.

WHITE CLOUD MOUNTAIN MINNOW

This is a rather romantically named pretty little shoaling fish that rarely exceeds a few inches in length. They can be easily bred in

Guppies can tolerate life in a coldwater tank, but must be introduced to lower temperatures over several hours, or even days.

The Zebra Danio makes an attractive addition to the coldwater tank.

the tank, particularly around springtime, where they spawn in fine-leaved plants as the first rays of morning sunshine warm the tank. To keep this fish happy, provide it with plenty of partial water changes. Ten per cent weekly would be ideal.

PLECOSTOMUS

These are actually from the area around the Amazon river, so strictly speaking they are tropical fish, but they do have a wide tolerance of temperature and will quite happily live in a tank with a temperature as low as 66 degrees F. They can grow quite large, sometimes exceeding 18 inches in length. They primarily feed on vegetable matter such as algae, and slices of potato. Additional food supplements such as tablet food should be provided to ensure that the plec receives a good balanced diet.

There have been a few new varieties of loaches appearing in the shops recently which have been wrongly called Chinese or Hong Kong plecs. These attractive fish come from the fast-flowing mountain streams of South East Asia and use their highly-compressed shape to cling to the rocks in the raging currents. Their specially developed mouth is perfect for rasping off algae from the rocks and roots they cling to.

Most of these new loaches do not exceed more than a few inches in length and make ideal additions to any coldwater tank.

SAILFIN SUCKER

This unusual fish (*Myxocyprinus asiaticus*) from China is a relatively new addition to the trade. It is believed to grow in excess of 24 inches, although very little is known about it. It is predominantly herbivorous, preferring to graze algae and soft plants than chase fish. It happily accepts tablet food, peas and other vegetable food. The huge dorsal fin is the characteristic item on this fish. On younger fish the fin is several times larger than the rest of the fish! They are peaceful by nature and, given adequate space and water quality, should present few problems.

NORTH AMERICAN SPECIES

These include some very attractive fish. Some are suitable for keeping in a tank, others grow far too large even for a pond. As these fish grow so well in cold water, there are current moves to license some species. This means that dealers need a licence to sell the fish and purchasers also need a licence to keep them. These moves have been fuelled by the successes of some of these fish after they have been inadvertently introduced to water courses around the UK. Species such as Channel cats, sunbass and lobster-like Crayfish sometimes have done so well as to be now considered a pest. Their thriving populations are endangering some of Britain's endemic species, causing extinction in some areas. If you choose to keep these fish, do not release them into the wild. You could do untold damage!

SUNBASS

All of these are predatory by nature and cannot be kept with small fish, so they are really for the larger tank. They are quite timid and do best if kept in small shoals among dense plant thickets. There are two main species available to the fishkeeper, the Pumpkinseed and the Blackbanded sunbass. Both have wonderful markings that rival even the most colourful tropical fish. A good supply of earthworms and frozen shrimp would suit these well.

ROSY RED, FATHEAD OR GOLDEN MINNOWS

These colour-sports of an American Minnow closely

resemble our native species, but are a pink-tinged gold in colour. As they breed easily and exhibit parental care, British-bred specimens are frequently available. They reach three inches. Male fish will defend a cave (or a fold in a pond-liner) while the young grow safe inside. An ideal aquarium fish.

DARTERS

A huge range of colourful minnow-like fish. Unfortunately not many are available for the fishkeepers in the UK but from time to time the odd one does turn up. They live in cool, fast-flowing streams and live on a wide range of small insect larvae. They adapt well to the aquarium and can even be spawned. They tend to lay their eggs under large, flat stones. To help them survive in the fast-flowing water they have adapted their anal fins to form a primitive sucker. This is stuck to the rocks and stones and allows the fish to hop around in search of food.

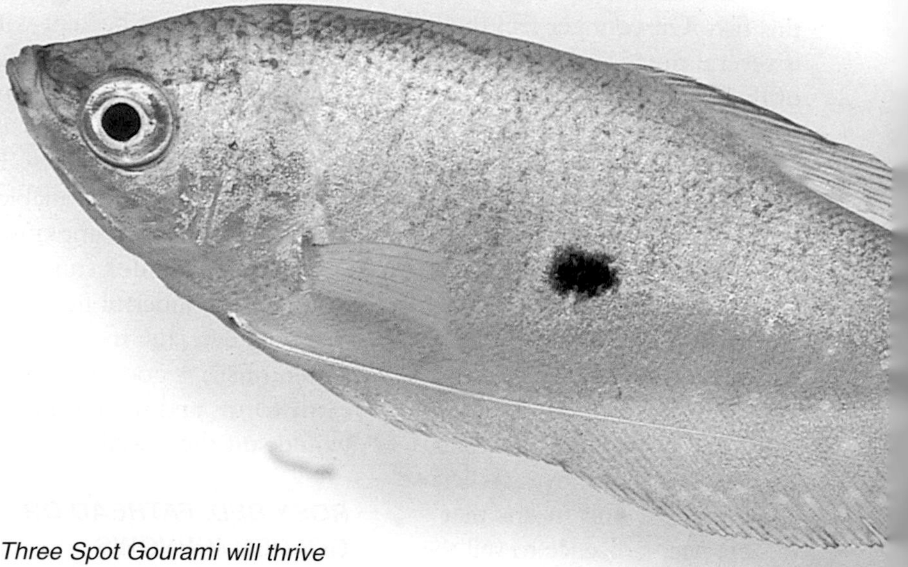

The Three Spot Gourami will thrive as long as the temperature does not fall below 65 degrees F.

BULLHEAD CATFISH

Also known as Channel Cats, these fish are very common but should really be avoided. They grow to as much as 3ft long and are voracious fish eaters. They appear to be very cute when small but quickly grow into deadly predators. If they must be kept it is strongly advisable that they are kept alone in a very large, well-filtered tank or pond.

FLORIDA FLAGFISH

With a bit of imagination the markings along the flank of the male fish resemble the Stars and Stripes flag of America. They are attractive fish which are ideal for a coldwater aquarium. They grow to a little over two inches in length and are easily bred. Prior to spawning, the male digs a pit in the gravel where he woos the female to lay her eggs. The eggs hatch and are ardently guarded by the male until they hatch. The male continues his protection until the fry are large enough to look after themselves. The adult fish accept a wide range of foods, from flake to live food, and will also tuck into any live plants you may happen to have. Occasionally these fish can be a little nippy, taking chunks out of any small fish which are not quick enough to swim away.

SHINER OR RAINBOW DACE

This is a very attractive fast-moving American species which is ideal for the right kind of coldwater community tank. They are shoaling fish by nature, being found in the fast-moving shallows in the wild. The males are distinguishable by their brilliant red/blue coloration and fantastic silver sides. Around spawning time the males develop a number of white tubercles around the head which signify their readiness to spawn. The colours are best seen when the tank receives a little natural light. The sight of a tank full of healthy Shiners with a shoal of Gudgeon is really quite spectacular.

2 Planning Your Aquarium

Before embarking on your new-found and fascinating career in fishkeeping, it is important to plan your tank carefully. Not only is this important for the well-being of your fish but it is vital that the tank fits into the overall decor of the room in which it will be kept. Aquaria should be seen as integral parts of the house, rather than just vessels in which your fish are kept.

TANK POSITIONING

Before buying your fish-tank, carefully consider where in your home it will be positioned. An area close to a door, for instance, should be avoided as the constant movement in front of the tank, and incessant vibrations from people walking past, will upset your fish to the point that they may be constantly diseased through stress. A position near to

Careful thought must be given to where you position your aquarium.

a sunny window should also be avoided, as the excess sunlight from the window will cause uncontrollable growths of algae and temperature fluctuations. Look for an area that is quiet, undisturbed and away from windows, or other heat sources such as radiators or fires. A dark corner of a room is ideal. It will help to brighten the room and will be easily visible from most places in the room. You obviously know your house, so position the tank where you think is best – just bear in mind the important points outlined above.

TANK CHOICE

Any visit to an aquatic shop will leave you quite awestruck at the range of aquaria now available. Whether you would like a triangular, hexagonal or column tank, they are now all available. The most important point to consider is whether the tank will hold enough water for the fish you would like to keep. Always remember that the fish you will be buying will mainly be juveniles capable of growing considerably larger. The species guide will help you to understand potential growth and sizes.

Cabinet aquaria have become far more popular over the last decade or so, and integrate into homes beautifully as a piece of furniture rather than just an aquarium. These are useful for storing away fishkeeping equipment in their cupboards and present a neat and tidy package. A cheaper version is the metal stand and aquarium. This comprises a simple metal stand, often capable of taking two fish-tanks, on top of which a sturdy sheet of wood is placed. Before the tank can be rested on the wood, a layer of polystyrene or cork tiles is laid down to cushion any irregularities in the wood which may crack the glass. This arrangement is functional but not very pleasing to the eye. If you choose to use an external power filter (discussed later) then there will be nowhere to hide it from view. Obviously the quality of the product is reflected in the price. Expect to pay more for a cabinet aquarium, especially those finished in real wood veneers, than you would for a metal stand.

There has been a recent upsurge in the popularity of acrylic tanks, mainly imported from the USA. Although perfectly functional, do not be tempted by the tall column tanks. These prove to be impossible to work with as they are often over six feet high and very narrow. Trying to clean the bottom of one of these tanks is not for the faint-hearted. If you do choose to buy an acrylic tank, stick with the conventional shapes – oblong, corner or even bow-fronted. These are perfectly suitable for keeping fish in without having to pull a muscle trying to service them. Be careful when cleaning the inside of acrylic tanks as they will scratch very easily; use nothing more abrasive than a sponge.

TANK VOLUMES

Here are a few simple equations to help you work out the volume of your tank. In most cases they have been simplified to 2 decimal places. For a large tank, that can make several gallons difference. The measurements are given in imperial UK gallons. Always check this, as some products refer to US gallons.

OBLONG TANK

Measure the length, depth and width of your tank in inches. Divide each of the figures by 12 to get the sizes in feet. Multiply them all together and multiply that figure by 6.23. The result is the volume of your tank in imperial gallons. To gain the volume in litres, multiply the gallonage by 4.5. The result is the volume in litres.

BOWFRONT TANK

Working out the volume can be a little difficult, but they are just a standard tank shape with an additional few inches at the front. With this in mind, calculate the volume for an oblong tank and add around five gallons. This is obviously a rough estimate but, unless you wish to delve into serious mathematics, it will have to do. To obtain the tank volume, measure the length, width and depth of the tank in inches and multiply them together to get the total cubic volume in inches. Divide this figure by 61.03 to give you the volume in litres. Divide this figure to give you the volume in gallons.

A rectangular tank and cabinet.

A Bowfront tank.

Example: a tank measuring 36 x 15 x 12 holds 23 gallons (106 litres).

36 x 15 x 12 = 6480 (volume in cubic inches)

6480 divided by 61.03 = 106.177 (volume in litres)

106.177 divided by 4.5 = 23.59 (volume in gallons).

For a bowfront tank add around five gallons to this figure to get the approximate volume.

CORNER TANK

This is another excellent way of fitting a tank into a small space as it can be tucked into a corner yet still manage to hold quite a quantity of water. Most units are quite tall, so are excellent for housing fish that require a little more height than a standard aquarium can offer. The only drawback I have found with them is lighting. Most fluorescent lights are a minimum of 12 inches long so they have to be positioned at the front of the tank. This obviously limits their suitability for certain types of set-up. Volume is simple to work out as, basically, a corner unit is a cube tank cut in half.

Example: A corner tank such as this measuring 30 ins along the back two panes and 24 ins high would contain 39.3 gallons (176.9 litres). Work out the volume for a

A Corner tank.

cube tank measuring the same as the dimensions on the corner tank.
30 x 24 x 30 = 21600 (volume in cubic inches)
21600 divided by 61.03 = 353.9 (volume in litres)
353.9 divided by 4.5 = 78.6 (volume in gallons of a cube tank)
78.6 divided by 2 = 39.3 (volume of the corner tank in gallons)

CUBE TANK

True cube tanks are those whose dimensions are exactly equal, length, depth and width being all the same. These can hold an enormous amount of water without taking up much space along a wall. They also apply themselves perfectly as freestanding tanks away from any walls or furniture. Due to their uniform shape, lighting can be fitted in very easily so they are ideal for planted tanks. As they have a large volume, ensure that your filtration is up to scratch.
Example: A 24 ins cube would hold 50.3 gallons (226.5 litres)
24 x 24 x 24 = 13824 (volume in cubic inches)
13824 divided by 61.03 = 226.5 (volume in litres)
226.5 divided by 4.5 = 50.3 (volume in gallons)

A Column tank.

COLUMN TANK

With the advent of acrylic tanks, these have become quite popular purely because they look different.
Maintenance is nigh on impossible in the larger units. Some are available which are up to six feet high – we have yet to meet anyone with seven foot long arms to reach the bottom of these units. Even with improvised equipment, caring for a tank like this is difficult. Unusual they may be, but I would advise you against buying one.
Example: A column 48 ins high and 12 ins across would

hold 19.8 gallons (88.96 litres). Using the formula πr^2h. π is equivalent to 3.142. The radius is half the diameter, so in this case the radius is 6.
So following the formulae;
3.142 x 36 x 48 = 5429.4 (volume in cubic inches)
5429.4 divided by 61.03 = 88.96 (volume in litres)
88.96 divided by 4.5 = 19.76 (volume in gallons)

L-SHAPED TANK

These basically consist of two rectangular tanks stuck together to form the shape. These are also very good tanks for keeping fish in, although getting them through doorways etc. can be difficult. To work out the volume, split the tanks into two separate units, work out the volumes of each one and add them together.
Example: An L-shaped unit with these measurements would hold 56 gallons (252 litres)
Tank 1) 21 x 15 x 18 = 5670 (volume in cubic inches)
5670 divided by 61.03 = 92.9 (volume in litres)
92.9 divided by 4.5 = 20.6 (volume in gallons)
Tank 2) 36 x 15 x 18 = 9720 (volume in cubic inches)
9720 divided by 61.03 = 159.26 (volume in litres)
159.26 divided by 4.5 = 35.39 (volume in gallons)
35.39 + 20.6 = 55.99 (total gallonage of L-shaped aquarium).

CONSTRUCTION

All aquaria designed to have a water depth greater than 17.5 inches must be constructed of at least 6mm glass. Aquaria 36 inches or longer should be of 6mm glass when the water depth exceeds 15 inches, and in 10mm glass where the water depth exceeds 18 inches.

Not all aquaria need to be sited on a base of polystyrene or similar material. Some tanks have what is known as a floating base where the bottom glass is suspended in a frame, thereby the only point of contact is the plastic frame. If you do need to site the tank on a cushioning layer, use cork tiles. These are neat and not as unsightly as polystyrene.

These volumes are for an empty tank, but do not forget to allow for 10 per cent displacement created by your rocks and gravel.

Equipment

When you have selected the type and size of tank to suit your requirements, the next step is to fit it with the equipment required for coldwater fishkeeping.

FILTRATION

This process keeps your fish alive by breaking down their waste into relatively harmless by-products. In essence, a filter is any surface through which water flows. On that surface, certain strains of bacteria flourish, which feed on the fish waste. As long as these are supplied with plenty of 'food' and oxygen, they will keep the water in your aquarium healthy enough to keep fish. Basically there are three types of filtration; mechanical, biological and chemical. A combination of the three provides the best filtration for your tank.

MECHANICAL FILTRATION

Simply, this is the physical removal of debris from the water. As water laden with dirt is passed through a mechanical filter, it is trapped by the fine mesh of the filter. Good examples of mechanical filters are filter floss, sponge and plastic mesh.

BIOLOGICAL FILTRATION

As previously mentioned, fish produce an amount of liquid waste into the water as a by-product of eating and breathing. This can be removed by certain strains of bacteria. These need a surface on which to live which allows a good through-put of water containing oxygen and food. Sponge is also an excellent base on which bacteria can live and breed. Other good options are the myriads of specific biological media currently available, such as the plastic 'bio' media and products such as ceramic tubes and sintered glass products. In order to work efficiently, biological media must be kept free of debris.

CHEMICAL FILTRATION

Certain chemicals may build up in your aquarium over a period of time which may cause problems with nuisance algae, discoloration of your water or even affect your fish's health. These can be removed by a partial water change, or their build-up can be reduced by a chemical filter media such as activated carbon or an impregnated foam. These types of filter media are readily available from aquatic stores but should be used sparingly. Activated carbon, for instance, is an excellent chemical filter media as it will absorb a variety of chemicals from the water. However, once saturated, it may release these chemicals back into the water in large doses which may be harmful to your fish. To prevent this from happening, change the chemical media regularly.

THE RANGE

There are numerous filters which embrace these principles, all of which are suitable for use. Let us take a look at some you may encounter.

UNDERGRAVEL FILTERS

These were popularised around thirty years ago by Graham Cox of Waterlife Research Ltd. They are probably the simplest form of biological filtration and encompass all of the principles of filtration talked about earlier. They comprise a corrugated and perforated plastic plate which fits across the entire base of the aquarium. In the back corner of the plate is an uplift tube which is roughly 1.5 inches in diameter and reaches up to the surface of the water. Once in place, your aquarium gravel is added to a depth of around 3 inches. The undergravel filter is generally powered by one of two means. The most common is the power-head. This is a simple 'motor unit' which fits on top of the uplift tube and sucks water up the tube. The water is drawn down through the gravel before being pumped back into the tank. On the grains of gravel certain strains of bacteria colonise and break down the fish waste to relatively harmless by-products, keeping the water free of ammonia and nitrite. The older and noisier method of running an undergravel filter used an air pump to power an airstone which was placed down the uplift pipe. The constant rising bubble causes the same kind of water movement generated by the power-head, but

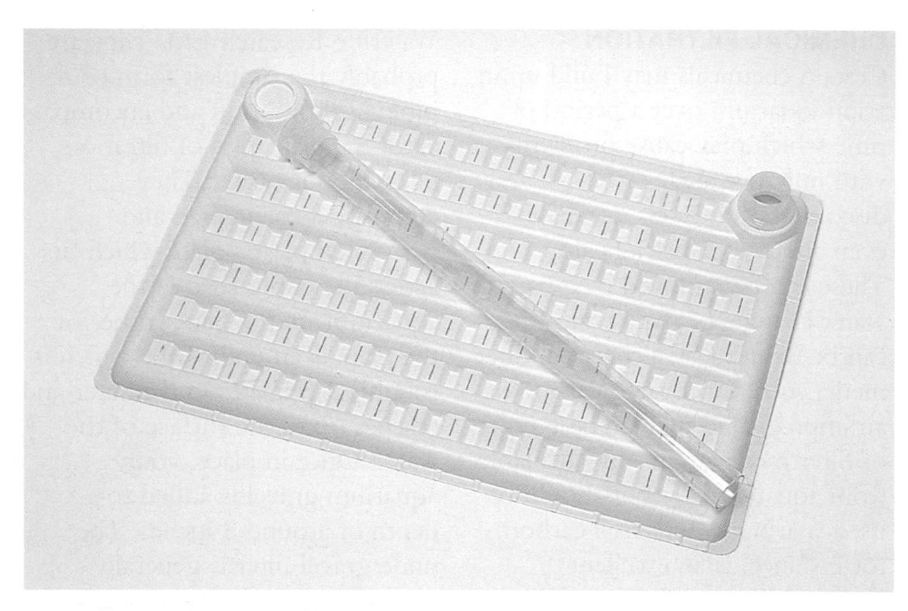

Undergravel filter plate and uplift

is a noisier and less effective method.

Though this is a very simple and effective method of filtration, there are one or two drawbacks. The constant movement of water through the gravel is unsuitable for a lush growth of plants (although there are species which will thrive, which we will cover later). Secondly, some fish delight in digging through the gravel, either in search of food or as a means of marking territory. As they excavate the filter bed, the water will take the easiest route through the plate, rendering the filter useless. There are methods of preventing this, using a gravel tidy – a thick plastic mesh buried into the gravel. Those problems aside, the undergravel is an excellent, cheap method of filtration which is easy to install and maintain.

INTERNAL POWER FILTERS

Most tanks now use these as their only means of filtration as they are relatively cheap and very easy to install and run. In essence they are made up of a sponge contained within a perforated plastic cylinder. Attached to the top of the cylinder is a power unit, very similar in appearance to a power-head. The whole unit is fixed to

Internal power filter.

the inside glass of your tank somewhere near to the surface, often in the back corner, by means of plastic suckers. The water is drawn through the slats at the bottom of the canister and through the sponge. Living in and on the sponge are millions of the same strains of bacteria which are found in an undergravel filter. As the water laden with fish waste is drawn through, they feed on it, converting it into less harmful products. These filters are mains-operated, using sealed, very low-wattage motors. They allow the prolific growth of plants, are cheap to run and maintain, and are readily available. They have few drawbacks, unless you intend to keep large or messy fish, in which case they are just not big enough to cope with the amount of waste produced. In this instance it is better to use a combination of filtration methods or a large external unit.

EXTERNAL POWER FILTERS

Imagine a large bucket that is watertight containing a selection of filter media. On top of the bucket is a power unit which returns the water to the tank. Basically, this is what an external canister filter consists of. They have room to hold a large amount of filter media, therefore they can deal with a greater load of waste. They are relatively large but are intended to be stored out of sight under the tank or behind it. The amount of equipment actually in the tank is minimal. In one corner of the tank, a pipe with a strainer on the end should be positioned. This is the outlet to the filter. At the opposite end, a spray bar is positioned at the surface of the tank to return the water. Although they cost considerably more than the other types of filtration discussed here, they provide an excellent means of filtering your tank. If finances allow, this type of filtration gives the greatest flexibility of all those mentioned here.

TRICKLE FILTERS

It is likely that you will hear of this type of filtration when visiting your local aquatic shops. Although not strictly applicable to a coldwater aquarium as discussed here, it may be necessary to know of them. Trickle filters work in a similar way to a municipal sewerage farm. Water is usually trickled through a rotating spray bar before passing down through some form of biological filter media. The filter media is suspended above the water surface in the trickle filter so that it has both an ample supply of water and of air. This combination allows it to work far more efficiently than conventional submerged filter media. Most trickle filter units are designed to fit under the tank, although recently there has been an upsurge in the number of compact internal trickle filter units. The standard under-tank affair is both costly and large and is usually used by dedicated fishkeepers with larger aquaria.

AIR PUMP

This simple piece of equipment is probably what revolutionised fishkeeping way back when it all started. Put simply, they are mini-compressors capable of pushing a volume of air along a narrow-bore flexible plastic pipe to a diffuser of some kind. There are numerous airstones, diffusers and air-operated ornaments available, and air pumps can be used to run basic filters. They are handy pieces of equipment to have on standby. In the summer when temperatures rise, oxygen levels drop. To help increase them, an air pump can be used to improve circulation in the tank. Air pumps also have the benefit of providing an attractive addition to any tank; the curtains of bubbles emitted from an airstone make a pleasing backdrop to your fish.

TEST KITS

These are an invaluable item of equipment, particularly during the initial maturing period of your tank. They are available to test a wide range of parameters in the

Air pump.

water but you should be concerned with just four at this stage; ammonia, nitrite, nitrate and pH. These help to highlight any problems which may appear in your tank and also indicate when it is safe to add your fish.

THERMOMETERS

These take many forms, but the best are the straightforward glass internal thermometers. These stick to the inside of the tank and give an indication of the temperature of the tank water. They should be positioned mid-way down in the tank. This again helps to indicate whether the temperature of the water is equal throughout.

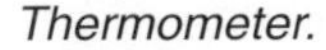

Thermometer.

LIGHTING

Lighting generally falls into two categories, that designed for good plant growth and that designed to enhance the colour of your fish. A single tube is all that is required if you simply wish to view your fish. If plants are to be grown, then closer consideration needs to be paid to the lighting system. The most common, and cheapest, form of aquarium lighting is the fluorescent tube. This consists of a tube and starter unit with waterproof end caps. The ballast unit within the starter is sealed in resin so that it is as waterproof as possible. Different wattage tubes require different power starters. The tubes available for aquaria have a multitude of spectrums, some of which enhance the colour of your fish and others which are specifically designed for plant growth. A combination of the two provides the best solution, giving your fish an excellent colour and promoting good plant growth. The quantity of light needed depends on the depth of the tank and, again, on whether you want to grow plants.

HOOD

Hoods are designed to house all of your lighting, to stop your fish from jumping out and the cat from getting in! They also help to tidy the whole tank up, making it an attractive piece of furniture rather than just a home for your pets. All cabinet aquaria come with a hood fitted as standard but separate all glass tanks and stands will need the hood buying separately. When looking for a hood, ensure that it is made from a non-corrosive material and that all its hinges and fixings are also made from a similarly inert material. Any wood hoods will need sealing with a good-quality varnish, and any hinges or handles must be made from stainless steel or brass.

DECOR

There is an Aladdin's cave of aquarium decor waiting for you to discover it. Once you start looking you will find something to satisfy everyone. Whether you fancy recreating the latest Disney film, or mimicking an Amazonian biotope, there is the correct gravel, rockwork and background to suit. This part of fishkeeping is very much down to personal preference and all that remains to be said is bear in mind the needs of the fish. Some fish require hard alkaline water, while others require soft acidic water. Some decor influences the chemistry of the water, so should be avoided unless being used for a specific type of set-up. Rock such as Tufa rocks make the water hard and alkaline, while bogwood will make the water acidic.

There is a simple test you can

Natural wood.

Choose rocks for aquaria with great care.

carry out to determine whether a rock is alkaline or not. Take your sample rock and drop a little weak acid on to it. If it begins to fizz or bubble it is alkaline. Failing that, take two samples of tap water. Measure them both into something like a sandwich box, or ice-cream tub, so that you have two separate samples. Measure the pH of both samples with your aquarium test kits and make a note of the readings. Add your sample of decor under test to one sample and leave the other sample empty. Cover both samples with a lid and leave for a week or so. Uncover and test the pH. The pH of the

Fantasy decor is not to everybody's taste.

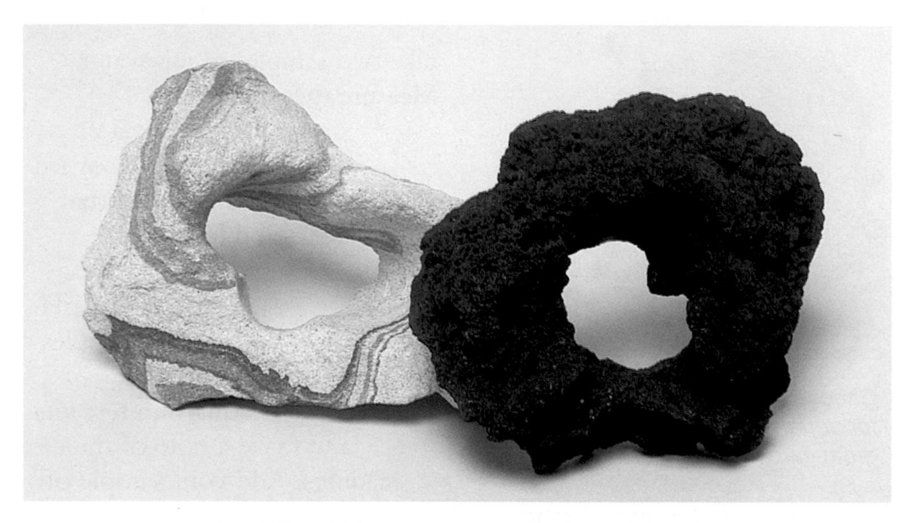

Many fish appreciate caves in rocks.

empty sample should be the same. The sample with the decor will have changed if the decor is either alkaline or acidic. When building the decor up, try to create caves, shaded areas and open areas for swimming. This combination should provide suitable conditions for most of the commonly-kept fish. Do not forget to leave room for planting.

PLANTS

These help to add colour, variety and interest to a tank. As they grow, they provide an ever-changing backdrop to your fish. A little care is needed, but if you choose plants that are easy to grow, you should encounter few problems. A handful of plants that are readily available and easy to grow include Hornwort, Elodea, Vallis, Hygrophila, Anubias,

Hygrophila

Vallisneria. *Amazon sword (Echinodorus sp).*

Cryptocoryne and Amazon sword. This is only a small sample – once you have 'cut your teeth' on these plants, why not try some others?

PLASTIC PLANTS

Not everyone wants to grow plants and, indeed, in some tanks, owing to their filtration, inhabitants and lighting, you would find it difficult to grow plants. Fortunately, all is not lost as there is an astonishing array of very realistic plastic plants available. If they are being used in a tank containing boisterous fish, it may be worth attaching the plastic plants to the decor with silicone so that they cannot be removed by fish tugging at them.

ADDITIONAL EQUIPMENT

Other useful pieces of equipment include a bucket which is only used for the fish-tank (detergents can kill your fish), a length of syphoning hose for water changes, a water filter (to make the tap water safe for your fish), a magnetic algae cleaner (for

Artificial plants may look reasonably realistic, but they lack the beneficial effects of live plants.

keeping the glass clean) and an old toothbrush (excellent for cleaning filters, rocks etc.). A few old towels are also useful for putting around the base of the tank when working on it.

YOUR SHOPPING LIST

Suitable tank, hood and stand.
Filter and relevant equipment.
Thermometer.
Gravel.
Decor (wood, rocks, plastic etc).
Plants (natural or plastic).
Background (internal or external).
Lighting equipment (tube, starter unit).
Test kits (ammonia, nitrite, nitrate, pH).
Dechlorinator.
Maturing agent.
Cable tidy or extension lead.
Bucket.
Syphon hose.
Algae magnet or scraper.
Net.

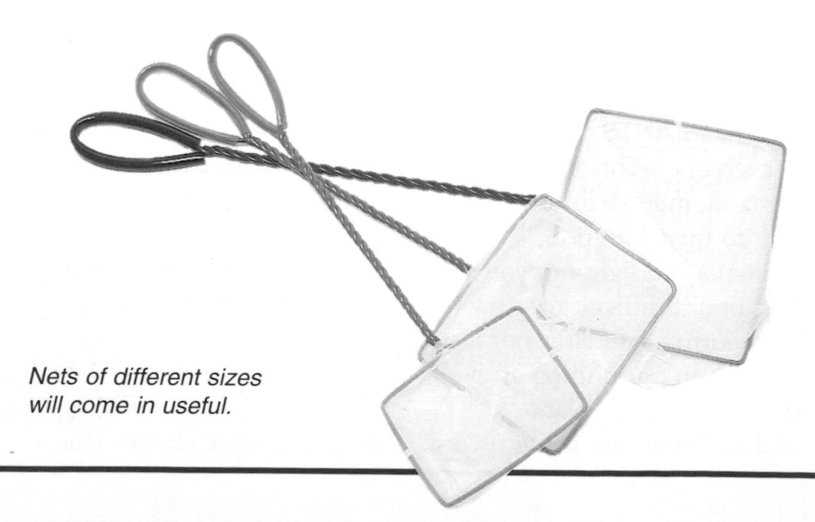

Nets of different sizes will come in useful.

4 Understanding Water Quality

Good water quality is paramount to healthy fish and, in fact, you should look at keeping your water rather than keeping your fish, as healthy water equals healthy fish. There are many factors which affect water quality, some of which we can control, and others which are out of our jurisdiction.

Regular tank maintenance goes a long way to keeping your tank in optimum condition. Even a simple tidy-up to remove any dead or decaying leaves can have a significant effect on the overall balance of your aquarium. This may sound absurd when most natural water courses are littered with tons of organic debris such as leaves and twigs, and various dead organisms such as insects and fish. What differs so greatly between a closed environment, such as tank, and an open one, such as a river or lake, is the massive bio-diversity. Open water courses are the permanent or temporary homes of literally thousands of creatures, both animal and vegetable. These interact with one another, the end result being a hospitable environment in which to live.

In a closed environment such as an aquarium, much of this life is missing and the 'biotope' is under the direct influence of the fish-keeper. Build-ups of organic debris, such as dead leaves, have to be physically removed. Stale water, rich in dissolved organic waste, needs to be changed and replaced with fresh water rich in nutrients to assist your fish and plants to thrive.

This may all sound incredibly complex but in practice there could not be anything more simple. Because of the closed nature of the aquarium, there are a number of parameters you must be aware of to ensure that your fish get the best possible care. All of these are easily tested for by inexpensive test kits

Water quality must be maintained to ensure that your fish remain healthy.

available from any aquatic retailer. Although water chemistry is important, do not feel that an in-depth knowledge of it is necessary to keep fish. It is far better to have a general understanding of the subject and a knowledge of the appropriate measures to take should the need arise. For ease of comprehension we will break the elements of water chemistry down into sections.

THE pH SCALE

This is probably one of the first things you will hear when broaching the subject of water chemistry. The pH scale simply measures the acid or alkaline content of your water. It is measured along a scale from 0 to 14. The lower the number the more acidic the water will be, therefore water with a pH value of 0 is very acidic, water with a value of 14 is very alkaline, while water at 7 is neutral.

Fish from all over the world live in vastly differing areas of water with equally diverse pH values. If these values had to be adhered to rigidly, a common coldwater community tank would be out of the question, as many of the fish would come from areas with different water quality. It is a tribute to the fishes' overall adaptability that they will thrive in a variety of conditions and, as long as the water is not wildly acidic or alkaline, your fish will suffer no ill fate. Many of the fish you buy will have already acclimatised to the local tap water values anyway.

There are a number of factors which affect pH, many of which will be dealt with later in the book. One of the biggest is carbon dioxide. This gas enters the water through a number of sources, most commonly through the respiration of plants and the breakdown of fish waste. During daylight hours, plants absorb carbon dioxide and respire oxygen, thus helping to provide excellent water conditions for your fish. Conversely, during the hours of darkness, the same plants take in oxygen and produce carbon dioxide, just as we do. In a balanced system, these processes do not present a problem, as the amount of carbon dioxide produced is quickly absorbed into the body of water in the tank. Only if the tank is under-filtered, over-stocked and over-planted does a problem arise. This is because the carbon dioxide combines with the water to form

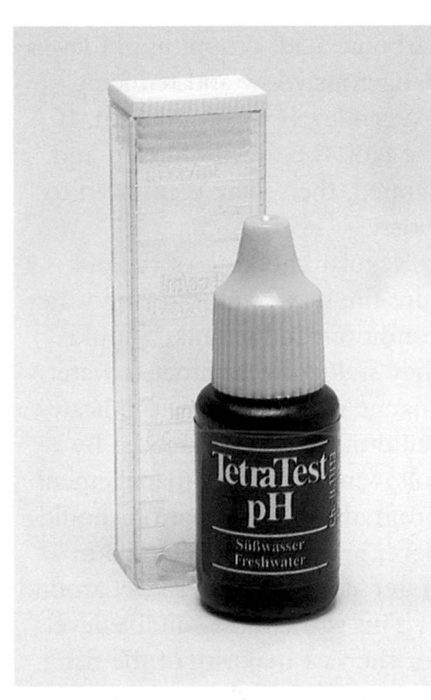

pH test kit.

carbonic acid, dropping pH levels dangerously low and reducing oxygen levels to a minimum. If the tank is correctly filtered and aerated, then there is no need to worry.

Regular checking of the pH is a sure-fire way of monitoring water conditions in the tank. Should they suddenly plummet, a water change is required, as it indicates a build-up of organic debris. By following the water management programme, the situation should never arise. Aim to keep your water at a neutral pH level around 7. This is the most suitable level for the vast majority of the fish you are likely to keep.

WATER HARDNESS

General (or Total) water hardness is a measure of the quantity of dissolved minerals in the water. Hard water has a high level of dissolved minerals, while soft water has few. These can be measured using a simple test kit. For general coldwater community fish aim for a level of around 10°dGH (179 ppm). Another term sometimes heard is carbonate hardness. This is a measure of the quantity of calcium carbonate dissolved in the water. Unless you are keeping marine (saltwater) fish, it is of little consequence, although do not exceed a level of 15°dKH (270ppm) in most tanks. In some areas, the tap water will be very hard. As a rule of thumb, if the pH is high, the hardness will be too.

THE NITROGEN CYCLE

This is probably the most important part of fishkeeping – the correct control and management of the nitrogen cycle. The fish we keep live in an artificial environment which is under your strict control. The waste produced by your fish cannot be washed away by the rains, just as the fish cannot swim away if the conditions do not suit it. You are in control, so you must make the living conditions for your fish as tolerable as possible. To do this you must understand the nitrogen cycle.

Fish produce ammonia as a by-product which is then excreted via the gills and the anus into the water. If not dealt with, it can build up to toxic levels, eventually killing your fish. Your filter, containing millions of bacteria, breaks down the ammonia into the less toxic nitrite, and then nitrate. The nitrate is used by the plants as a fertiliser; any remaining nitrate can then be diluted with every water change carried out by the fishkeeper. The whole process takes a month or two to establish initially but will survive a lifetime if properly cared for. Let us take a look at it in greater detail.

AMMONIA

This is the first product in the nitrogen cycle and also the most deadly. The aquarium contains a vast amount of proteins, in the form of plant material and foods, introduced to satisfy the needs of your fish. On the whole, proteins contain a large amount of nitrogen, hydrogen, carbon and oxygen and it is from these compounds that ammonia is formed. Due to the metabolism within the fish and the actions of certain strains of bacteria, the proteins are broken down to form ammonia by joining the nitrogen and the hydrogen molecules to form NH4 (ammonia). The remaining carbon and oxygen form carbon dioxide, which leaves the water in the form of minuscule bubbles, or forms carbonic acid.

The ammonia enters the water from the fish in the form of urea, uric acid and amino acids. These, in turn, are used by bacteria, known as heterotrophic bacteria, as a source of food. As they are used, they are slowly broken down

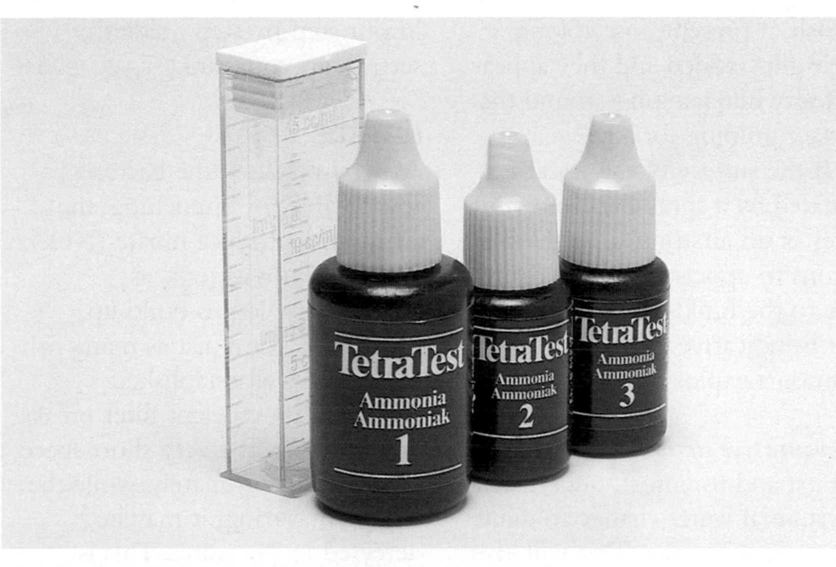

Ammonia test kit.

into less harmful products such as nitrite and nitrate. Due to its chemical nature, ammonia is harmful to aquatic life in relatively small doses. Its toxicity also increases with temperature and pH. Therefore, a tank with hard water and a high temperature is particularly susceptible to the effects of ammonia. As mentioned previously, heterotrophic bacteria use ammonia as a source of food and, in liberating the energy locked within it, reduce it to a simpler compound known as nitrite. This is the next step in the nitrogen cycle.

Signs of adverse ammonia levels
i) Fish, if present, lose colour, their gills redden and they appear listless while hanging around the surface gulping for air.
ii) If the surface of the water is agitated by a spray bar from your filter or an airstone, a 'froth' begins to appear. This is actually due to the build-up of proteins but is indicative of an impending ammonia explosion.

Preventative action
i) First and foremost, undertake a substantial water change to dilute the toxic ammonia. This will give your fish instant relief.
ii) Check that your filter is working correctly. In an established tank this is a common cause of sudden ammonia blooms, as the waste is no longer being dealt with. In a new tank, the bacteria needed to break the ammonia down will not have established in sufficient quantities to deal satisfactorily with the waste. In such a scenario, a shop-bought cure is required which chemically locks away the ammonia, making the water safe. Once safe, add a rapid-maturing agent to the filter to speed up the period needed to establish the filter. Maturing your filter is important and will be dealt with in our step-by-step guide to setting up your tank.

NITRITE

As a by-product, the bacteria responsible for 'munching' the ammonia, produce nitrite (NO_2). Although not as toxic as ammonia, if left to build up, nitrite can claim just as many fish lives. In a well-established aquarium an efficient filter breaks down nitrite in a very short space of time. Unfortunately, while the tank is maturing, it may be detected in the water. This is because of the time taken for a species of bacteria known as *Nitrosomonas* to establish. These bacteria effectively reduce the nitrite into a far less toxic product known as nitrate.

Signs of adverse nitrite levels
As it is as toxic as ammonia, these do not vary a great deal from those previously mentioned. Many 'unexplained' deaths can be attributed to the presence of nitrite. Fish which were perfectly happy yesterday are dead today, showing no symptoms except reddened gills and an opaque or dull coloration due to the build-up of mucus over the fish's entire body.

Preventative action
First course of action is a water change. This will rapidly dilute the nitrite present to a far less toxic level and will grant your fish a momentary reprieve from the problem. Next step is to prevent it from happening again. Check that your filter is large enough to cope with the number of fish that your tank holds. This can simply be found out by looking at its box or details on the motor unit. Most modern filters have very clear instructions for their use and suitability for the type of fish that you choose to keep. Stop feeding the fish for a few days, no food going in means no waste being produced. Some nitrite-reducing agents are available from your local aquatic shops which also add useful bacteria to your filter, thus helping to prevent a nitrite problem again.

NITRATE
Of the products of the nitrogen cycle, nitrate is probably the least harmful. It is readily taken up and used by plants as a form of nourishment and quite a high level is needed before your fish show

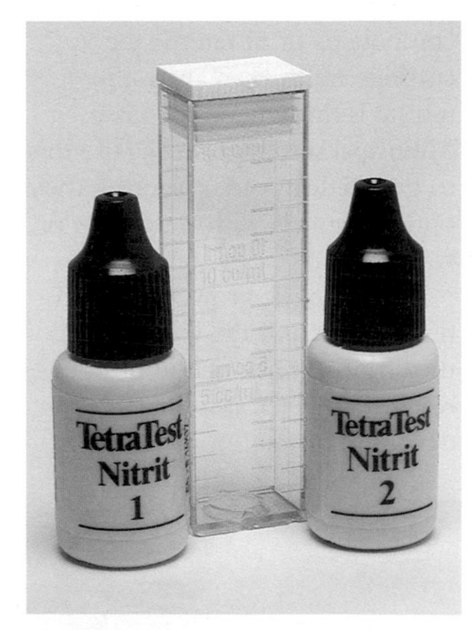

Nitrite test kit.

any noticeably ill effects. Nitrate is produced by *Nitrobacter* bacteria which use nitrite as a form of food. As the nitrite is broken down, nitrate is produced. Nitrate can be filtered out of the water but very specific conditions need to be set up before this process will take place.

In some filters, particularly undergravels and external canister filters, this process takes place inadvertently without the fishkeeper's intervention. Normally nitrate is removed by water changes and the prolific growth of plants. As the plants are pruned, the cuttings taken contain compounds of nitrate. If the cuttings are thrown away, the nitrate is thrown away with it. Municipal sewerage works use this to good effect, often planting their outlet channels with watercress or water mint.

These prolific growers take up nitrate very quickly; by the time the effluent, once rich in nitrates reaches the river, it is almost devoid of nitrate.

Signs of adverse nitrate levels

Nitrate never directly kills any fish, but it slowly reduces the fish's immune system to such an extent that the fish succumbs to a very bad dose of a disease and dies. If your fish are continually ill with one disease or another, or they take a long time to respond to treatment, chances are that you have a high reading of nitrate. Extensive growths of unsightly algae are another sign of high nitrate. The slime algaes seem to thrive in less than favourable conditions, particularly the blue-green and red slime algaes.

Preventative action

Nitrate cannot be avoided as it is the natural end product of your filtration system. It is important to understand that it will always be produced as long as your filter is working properly. In effect, nitrate can only be controlled. Regular partial water changes, good plant growth and good general tank maintenance all help to establish a tank with little or no nitrate.

5 Setting Up And Maintenance

This is a step-by-step guide to establishing your aquarium, and the on-going care that is required once your tank is established.

POSITIONING THE TANK

Position the stand or cabinet in a level, quiet position away from direct sunlight. Be sure that there is at least three or four inches between the wall and the tank to allow the wires and pipes to pass down without any trouble. Place the tank on to the stand, ensuring that a layer of polystyrene or cork tiles are placed on the stand first if the tank needs them. Once you are happy with the position, wipe the inside of the tank with a clean, detergent-free cloth and remove any stickers. The damp cloth will remove any dust which may have collected.

FITTING THE BACKGROUND

Before filling your tank or positioning any equipment, you will need to stick your background in place. Cut it to size, and stick it to the outside of the tank with tape. Some newer backgrounds fix inside the tank, so check when you buy it.

THE SUBSTRATE

After washing your gravel, place it carefully into the tank aiming to achieve a gentle slope from the back of the tank to the front. This helps any solid debris to fall to the front of the tank where it can easily be syphoned out. If you have chosen to use an undergravel filter, install this before the gravel is added.

THE FILTER

Now the gravel is in place, position your filter. Try and site this at the end of the tank nearest the plug to save unsightly leads from trailing everywhere.

THE DECOR

With the hardware in place, the decor can be added after a thorough washing. Try and hide the

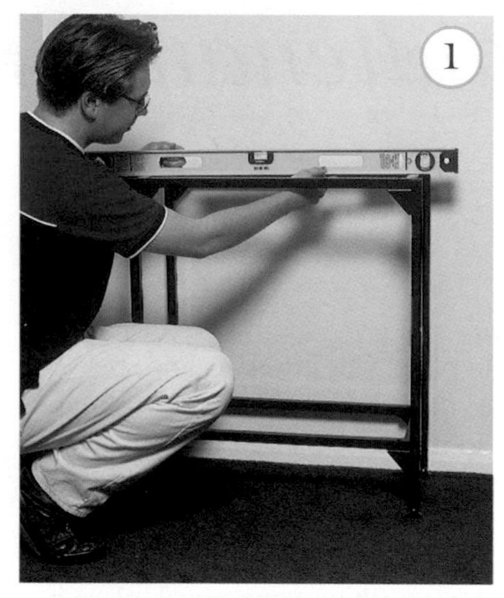

Positioning the stand.

Polystyrene sheets are fitted on top of the stand.

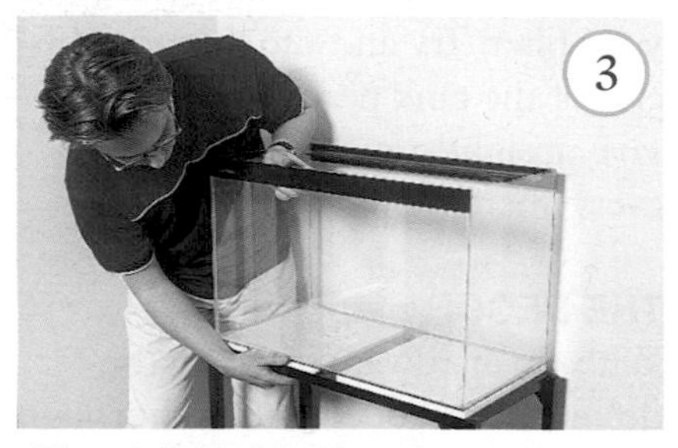

The tank is placed on the stand.

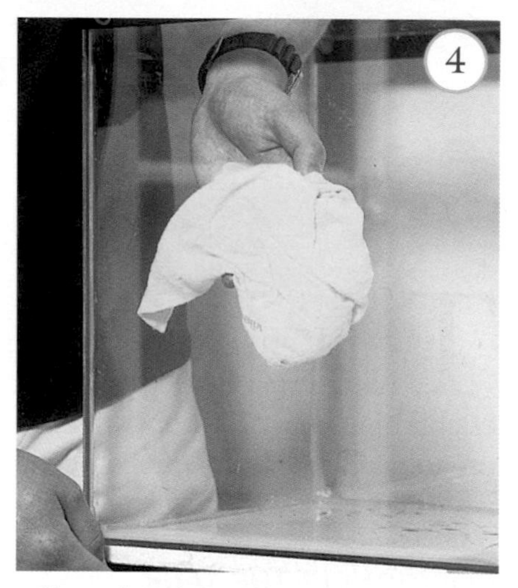

Clean the glass with a detergent-free cloth.

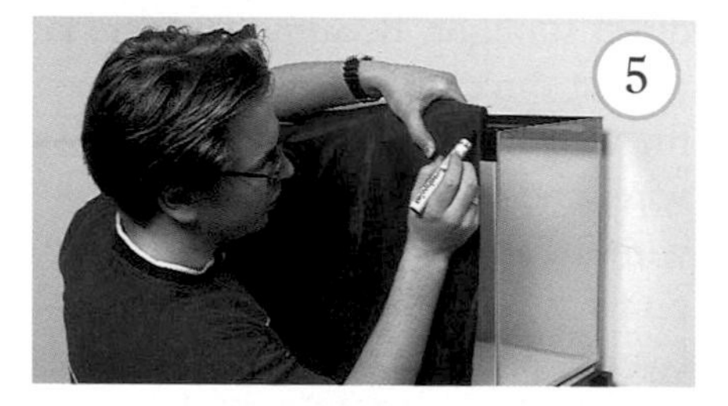

Cut the background to size.

Fix the background in place.

The gravel must be washed before it goes into the tank.

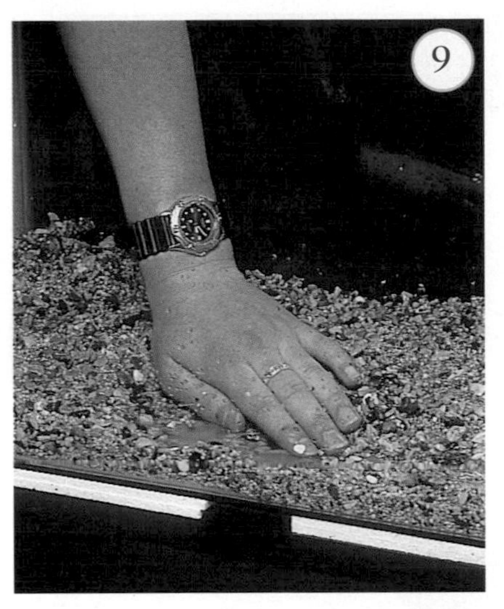

The gravel should be sloped from the back of the tank to the front.

Pour the gravel into the tank.

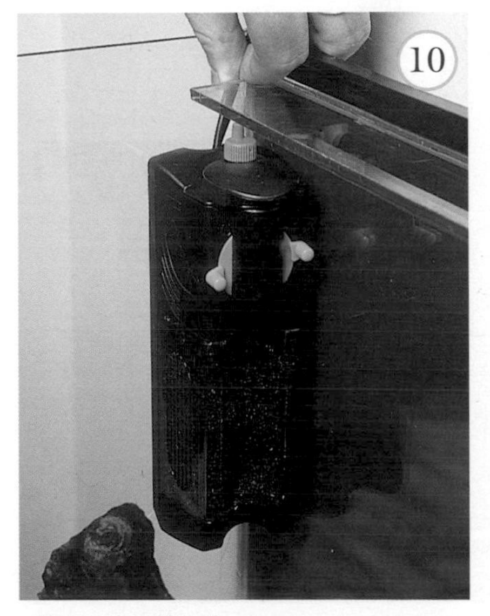

Fixing the filter in place.

The rocks and any other ornaments are positioned to create hidey-holes for the fish.

Time to fill up the tank.

The lighting is fitted to the hood of the tank.

The tank is ready for planting.

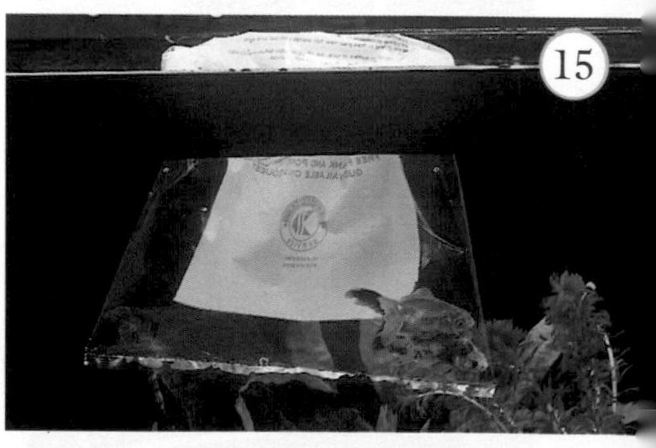

Allow the temperature in the bag and the tank to equalise before releasing the fish.

The finished set-up.

equipment and provide plenty of caves and hidey-holes for your fish, but do not forget to leave some open areas for swimming and for planting up. If using a lot of rock work which extends to the surface of the tank, consider fixing the rocks together with silicone so that they do not fall down. This should be done a day or so before setting the tank up. Any rock or woodwork should be pushed well into the gravel for support.

FILLING UP

Once you are happy with the decor arrangements, it is time to fill the tank. Place a saucer or small plate on to the gravel and pour the water onto this. This stops the gravel from being disturbed and the rocks collapsing. Ideally the water should be lukewarm although cold water will be fine. If you choose to use lukewarm water, heat it up by boiling it. The hot water from the tap may contain copper which is harmful to your fish.

As the water fills the tank, add the required amount of shop-bought de-chlorinator to prevent damage to your fishes gills. Fill the tank half-full.

PLANTING

Time to roll up your sleeves and delve into your fish-tank for the first time. Your plants need to be planted and it is easier to do it with a half-empty tank than a full one. Before filling your tank, it is worth drawing up a quick sketch of where you would like to position the plants. Try and group the same plants together so that you have thickets of the same species. This looks far more effective than bits and pieces thrown together. This also applies to plastic plants. When finished, fill your tank to within an inch or two of the top.

SWITCHING ON

Now that all the water, decor and plants are in, it is time to switch everything on. You will have around four or five plugs coming out of your aquarium and you have the choice of either plugging them into an extension lead or using a shop-bought aquarium cable tidy. The cable tidy gives you a neat solution to 'cable spaghetti' with two switches to control your lights and your filter. Before switching everything on, stick your thermometer on to the tank at the opposite end to the filter.

Everything should fire up first time, although, if you are using an external canister filter, it may need priming. The fluorescent tube needs to have the end caps pushed on firmly and then to be clipped into the hood. To help the filter to mature, add the required dose of maturing agent. Most maturing agents need to be added on a daily basis; be sure to read the instructions carefully.

MATURING THE TANK

The following day, check that the filter is still working and that the temperature showing on the thermometer is what you anticipated. Any work carried out in the tank should only be done with the power switched off. Expect the tank to look a little cloudy for a week or so but this will soon clear as the tank settles down. After two or three days, check the ammonia and nitrite. If the maturing agent is doing as it should, you should have a slight ammonia reading and possibly a nitrite reading too. This is a sign that your filter is beginning to mature.

You cannot keep any fish until these readings have returned to normal. Expect to be able to keep fish within a week or two but it can vary. It really is important that you are patient throughout this time as rushing things could be disastrous. When the amonia and nitrite reading is zero, it is safe to add one or two small fish. Again, monitor the nitrite over a few days. If it remains stable, another one or two small fish can be added. The secret to successful fishkeeping is patience and stability. By keeping the water conditions stable, your fish thrive. Do not do anything drastic like 100 per cent water changes. Everything should be done gradually, a 10 per cent water change weekly is far more beneficial than a 50 per cent change once a month.

FEEDING YOUR FISH

Like you and I, fish require a good, healthy, balanced diet. It should have the required amount

Fish require a well-balanced diet.

of proteins, fats, carbohydrates, vitamins and minerals and include things like roughage. Most shop-bought dried foods such as flake, tablet or pellet foods can provide this but nothing beats a bit of variety. For the best results from your fish, aim to feed a mixture of basic dried food, frozen and live food. To help you understand their benefits, each type is described here in detail.

DRIED FOODS

These include flake food, tablet, pellet, granular and freeze-dried live foods. Flake food is a combination of fishmeal, vegetable protein and all the vitamins, minerals and roughage that your fish may need. It is mixed into a paste and then passed through a series of hot rollers to form a cooked sheet which is broken up to form the flake food, looking very much like a small cornflake. Many fish are reared on these and, indeed, most aquatic shops feed little else. Tablet food is a compressed pill of flake food which sinks to the bottom of the tank where your catfish can reach it. Pellet and granular food is very similar to flake, containing all the nourishment that your fish are likely to need. As they are in a pellet form they tend to float longer and are ideal for feeding larger, messy fish. Granules are the same as pellets but, being a little smaller, are ideal for the medium-sized fish in your tank. Freeze-dried food is different altogether, as it is actually the freeze-dried bodies of various live food items such as tubifex, daphnia and shrimp. These have a variable protein content but are excellent at adding roughage to the diet.

Dried food is available in a variety of different forms.

LIVE FOODS

These can include anything from a household spider to specifically-collected aquatic creepy crawlies. They are an excellent source of food particularly for conditioning your fish ready for spawning.

When feeding aquatic live foods be careful not to introduce undesirable insects such as dragon fly larvae and damsel fly larvae. These are extremely predatory and will snaffle any unsuspecting small fish. Simply study the live food before using it. All live food should be washed before using it to ensure that nothing is added to the tank that could be harmful. Below is a list of live foods and how they should be fed and prepared.

Daphnia: The humble waterflea. A first-rate live food. Check while in the bag for interlopers such as larvae, leeches etc. Before adding to the tank, rinse under the tap in a fine mesh net, or the toe end of a pair of tights, which makes a good filter.

Bloodworm: The larvae of a gnat. Superb conditioning food. Do not feed to very small fish, as has been known to eat its way out again. Rinse under the tap in a fine mesh net.

Tubifex worms: Another good conditioning food. Can be very smelly, best prepared by placing in a bucket under a gently running or dripping tap. As the water overflows, the new water cleanses the worms. May take a day or two before ready but well worth it. Ideally feed through a conical worm feeder.

Conical worm feeder.

Brineshrimp: This tiny saltwater shrimp is packed with protein, hence it is good for growing fish. Available from most aquatic shops, needs gentle rinse before adding to tank.

Rivershrimp: Collected from the brackish creeks and dykes around coastal Britain, the rivershrimp is an excellent food for larger fish. Rinse before use.

Earthworms: Cheap, easily available and packed with goodness, earthworms are superb food for any fish. Before using, place them in a ventilated tub with moss. This allows them to empty their guts full of phosphate and nitrate-rich mud. Smaller fish may need them chopping up.

TANK MAINTENANCE

In order to have a trouble-free tank, a few basic jobs need to be carried out on a regular basis. Some of the smaller jobs need to be daily, while other more time-consuming tasks may only need monthly attention. It is a very good idea to get into a regular regime for looking after your tank. By doing this, you will be diluting any problems that may arise by catching them at an early stage, and your fish will suffer very few, if any, problems.

Most problems that fish encounter are often the direct result of neglect. Earlier in the book we talked about the fish being completely in your care as they are confined to such a small area. The actual area that the fish live in is often inconsequential as long as the quality of water is maintained at a high standard.

You will also know from earlier parts of the book that water quality can be affected by seemingly insignificant occurrences. A build-up of a couple of dead plant leaves may seem of little concern, but their eventual bio-degeneration adds a load to your filter, which may be all that it takes to push it over the edge, and poor water quality results. Regular maintenance sees right around this problem, as debris is never allowed to build up to a stage where it may be a problem.

The following timescales may be relaxed slightly as your experience grows. Although you are unable to communicate with your fish directly, you will begin to establish an almost innate sense about their well-being. At a glance you will be able to tell whether they are behaving correctly or not, whether they appear stressed or unhappy, and you will know exactly what to do to correct the problem. That aside, follow the guidelines set out here until you fully understand the needs of your fish.

DAILY TASKS

While admiring your magnificent aquarium, give the fish a good check over. Do they appear healthy and bright? Are they all there? Are their eyes clouded over or is their colour a bit lacklustre? If so, this could be a sign that things are not as they should be in your tank. Ammonia could be building up or your fish may have contracted a slight bacterial or parasitic infection. Often these symptoms are accompanied by flicking and scratching, which shows that the problem could be quite advanced. If this is the case, consult the disease section in this book, or the water quality section, for further ideas about how to sort the problem out.

While looking at the fish, take a quick glance at your thermometer. Is it reading a sensible temperature? Although your tank is not heated up directly, it receives warmth from its surroundings, especially the lighting unit used to illuminate the tank. In summer, the tank can get very hot, which will cause low oxygen levels. This can be corrected by increasing the amount of aeration that the tank receives and carrying out a water change with slightly cooler water. In hot tanks, your fish will appear much more active than usual and will often be seen hanging around the surface gasping for air.

While checking the temperature, have a look at your filter. Is everything working OK there? Does the inlet looked blocked with leaves or other tank 'rubbish'? If the filter is unable to work at full power, waste may be allowed to build up to dangerous levels which could be damaging to your fish.

The most enjoyable daily job is, of course, feeding your fish. With larger coldwater fish it is quite easy to get them hand-tame, where they will eat food directly from your fingers. Feeding should take place at least twice a day; smaller, more frequent feeds always being better than a single large feed. You should be feeding a good varied diet, so try and vary the food each time you feed your fish. Just as you and I would not appreciate porridge for every meal, neither will your fish. If you can manage to feed your fish more often than twice a day do so. Several very small feeds are better still. However, for one day a week, it is not a bad idea actually to skip the food altogether. This stops your fish becoming unattractive, overfed barrels and helps to keep the tank free of snails, and uneaten food – of which there should not be any anyway – as the fish are actively encouraged to search around for anything edible.

WEEKLY TASKS

As with all areas of fishkeeping, small adjustments are tolerated much more successfully by your fish than big ones. With this in

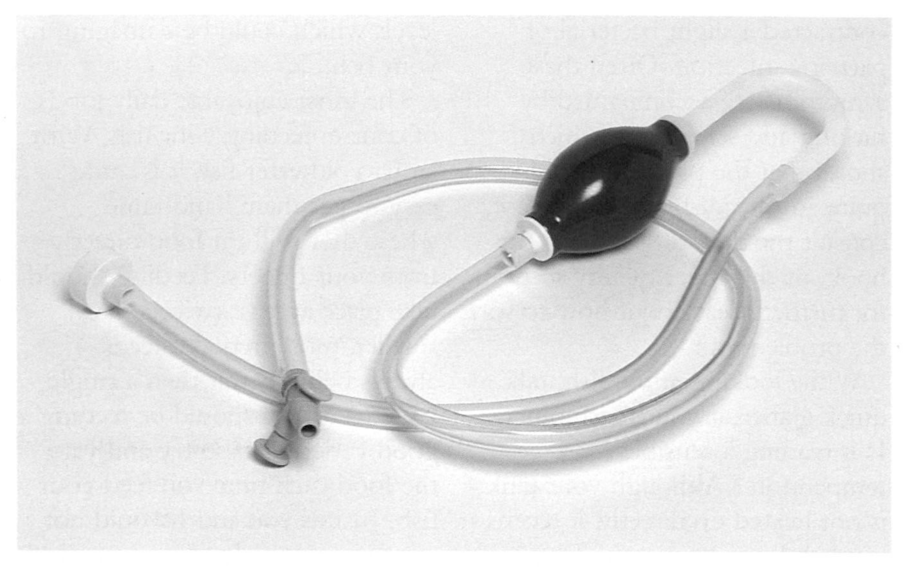

A syphon hose is required for changing the water.

mind, small weekly water changes are much better than a large monthly change. It is recommended that you carry out a water change of between 10 to 15 per cent of the tank volume. With tanks up to 36 inches in length this is little more than a bucketful a week. Not only do your fish benefit from this gentle care, but it is also a lot less time-consuming than a monthly large change.

Removing the water is a doddle. Using a syphon hose, the water is simply syphoned out into a bucket until sufficient has been removed. While removing the water, try and suck up as much debris as possible. There are specific gravel-cleaning syphons available, which are very useful. The replacement water needs to be treated with a dechlorinator before it goes into the tank. This removes the toxic metal and chlorine that would otherwise damage the gills on your fish.

Although your fish are known as 'coldwater' they would not appreciate a bucketful of freezing cold water straight from the tap being poured into their tank. Three-quarters fill your dedicated water-change bucket, and then top it up with water from a boiled kettle. Do not use water from the hot tap as it will be laden with metals picked up from the water tank and pipework. Using the tried and proven 'elbow test', bring the water to roughly the same temperature as that in the main tank. When happy that the water is roughly the same temperature, gently pour it into the tank, trying not to disturb your gravel or other decor which is there.

Before carrying out the water change, prune your plants to remove any dead or dying leaves which may break off and clog the inlet to your filter. This can be done using a sharp pinch from your nails or a clean pair of scissors. Not only are you removing unattractive dead leaves, you are also taking away leaves that are nitrate and phosphate-laden and would otherwise add to the load on your filter as they broke down.

If your lighting has been on for the required amount of time (12 hours a day) then the chances are you will have a slight build-up of algae on the inside of your glass. This should also be removed

Equipment for cleaning tank glasses.

before carrying out your water change. It should scrape off easily using either an algae scraper, a magnet scraper or an abrasive pad designed specifically for the job. It will temporarily cloud your water but will soon clear as it is syphoned out, and the filter deals with the rest. Do not clear the algae from the back glass as the fish will no doubt enjoy grazing from it, in addition to their diet.

After the tank has cleared, it is a good time to check your filter media to see if it needs lightly rinsing in the old tank water. Your mechanical or polishing filter media will probably need a quick rinse and, after six months or so, some of your biological media will also need a rinse through.

FORTNIGHTLY TASKS

This is the time to bring out the test kits for their regular fortnightly tank-water check. In the initial settling-in period of a few months, it is advisable to check nitrite and nitrate as well as pH. This should give you a very clear picture of what is going on in the tank. It is advisable that a clear record is kept so that any trends can be spotted easily.

A magnetic pad is used to eliminate the build-up of algae.

Should a high nitrite level be noticed over the usual settling-in period of a month or two, consider whether you are overstocked, overfeeding or the tank is underfiltered. All could cause high nitrite and all are easily solved. Nitrate is a compound that will unfortunately build up in most systems and is only easily cleared by a water change. Due to your weekly water change regime, high nitrate is unlikely to be a problem unless your tap water already has a high residual reading.

No matter how careful you are, it is inevitable that some flake food will be spilt on the cover glasses or condensation tray between your lights and the water. Now is a good time to remove it, and any limescale build-up, simply by wiping over with a cloth soaked in spirit vinegar. This is a harmless product to your fish, but it is advisable that the procedure is carried out over a sink and the tray or glasses rinsed thoroughly before being put back into position. Doing this ensures that the maximum amount of light reaches the tank, which is beneficial both to your plants and for your viewing.

MONTHLY TASKS

Hopefully your plants will be growing like mad and need pruning back to maintain a good bushy growth. Now is the time to do it. Using sharp nails or scissors again, remove the extra growth just above a leaf node (depending on the species) and either discard it or use it to further colonise the tank.

The media in your filter will probably require a rinse through with tank-water so as to preserve a healthy colony of bacteria. Undergravel filters will need going through with a gravel cleaner, paying particular attention to the areas of the filter next to the uplift tubes, as this is where most of the dirt and debris will have collected.

If any of the decor in your tank can be moved without utterly ruining the look of the tank, try and move them just so that you can get the gravel cleaned under there as well. You will be amazed at quite how much debris does build up under the decor in a tank filtered by an undergravel filter. Should you be using any form of chemical filter media, it would be prudent to consider changing it for new, or recharging it if possible. Some chemical media are unable to 'store' what they have absorbed; these can wreak havoc in a tank should they choose to release their chemicals back into the water. To avoid this, regular monthly renewal is the only option.

This may all sound like a workload from hell; in actual fact, most of it will become second nature after a few attempts and it really is only a case of answering your fishes' needs. They will be the first to indicate a problem should one arise in the tank. Daily watching and appreciation of what they are doing will help you decide which is the most appropriate course of action to take. Although your fish cannot speak, they have an amazing way of communicating.

Your fish will thrive in a well-maintained tank.

6 Special Considerations

Once your aquarium is firmly established it can be very satisfying to breed and raise some of your own fish. To breed your fish to any level of success, a separate breeding tank is needed. For most fish, this need not be massive, often a tank measuring 18 x 12 x 12 ins is ample. As long as the tank is filtered and well cared for, you should encounter no problems. Most eggs and fry will not withstand a battering from a power filter; an air-driven sponge filter is ideal. Set the tank up somewhere quiet away from any disturbance. Some of the more difficult fish need peace and quiet in order to breed.

EGG SCATTERERS

Most coldwater fish spawn in groups or in a pair. Spawning takes place after a lengthy courtship, involving the males displaying superb coloration and impressive courtship dances. Male goldfish in particular develop white bumps, known as spawning tubercles, all over the head and pectoral fins – the male with the best colour and display being the strongest and the one most likely to breed with the female. Often egg scatterers require certain triggers to start spawning, such as cooler water, shorter daylight hours or different water chemistry. Before attempting to spawn any egg layer, read up on that fish's particular needs.

A suitable breeding tank for egg scatterers would be set up with a layer of marbles on the bottom and a clump of fine-leaved plants, or spawning mops, dotted around the tank. Introduce the fish in small groups and try and keep the males and females separate on either side of a divider. Feed the fish on plenty of live food until both males and females are plump, then introduce them to each other just before switching off the lights.

Breeding is a fascinating aspect of fishkeeping.

Often these fish will spawn at first light. Check the spawning mops for the eggs, which will resemble tiny amber-coloured beads. If some are present, remove the fish back to the main tank. The fry should hatch after 3 to 5 days. If the eggs turn white, they are infertile and should be carefully removed. The newly hatched fry will live on the contents of their yolk sacs for the first few days and then will eat very small live food such as newly hatched brineshrimp and infusorians (water containing microscopic plankton). As they grow, increase the size of the food until they are taking flake.

BUBBLE NEST BUILDERS

An unusual 'coldwater' fish, the Paradise fish, is a bubble nest builder. These hail from China where the temperature can fall quite low, often down as far as

65 degrees F (18 degrees C). Their fascinating breeding habits make them popular fish for indoor coldwater aquaria. Breeding is quite an elaborate process, with the male spending a huge amount of time constructing a nest of mucus-covered bubbles interlaced with pieces of plant material.

After a courtship period, the pair of fish meet up under the nest before they mate. The fertilised eggs are collected by the male and spat into the bubble nest one at a time. When finished, the male guards the nest until the eggs hatch. If any eggs fall out, the male spits them back in again. As the fish breeding in this way are labyrinth fish, meaning that they have a primitive lung, the fry need to breathe warm atmospheric air as soon as they are born. To achieve this, the water level in the tank should be dropped and the cover glasses should be kept firmly shut.

A tank for breeding bubblenesters should be quiet, with a temperature around 75 degrees F (24 degrees C). The tank should be planted out with plenty of vegetation, including floating plants, and the tank level should be low, half-filled is fine. The fish, females first, should be added to the tank and fed well on live food. This conditions both fish ready for spawning. Often the male can be aggressive to the female, so plenty of cover should be provided for her to take refuge in, should she need it. After pursuing the female, the male 'dances' under the nest to entice the female to spawn. When ready, the pair embrace under the nest until eggs are produced. They are a truly amazing

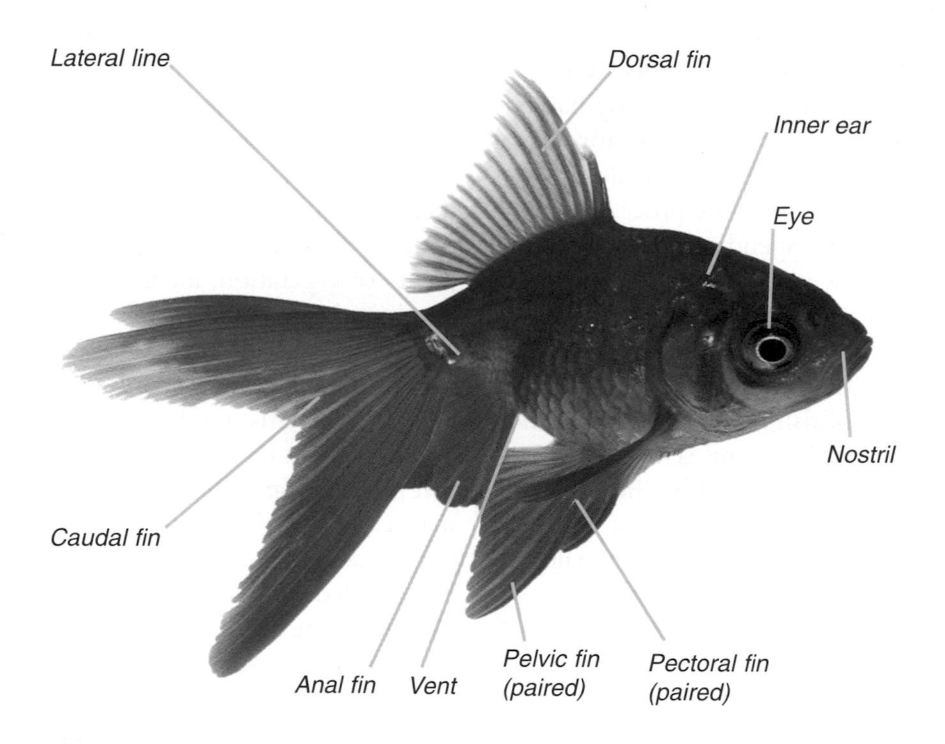

Although not essential to have a thorough knowledge of your fish's anatomy, it does help in diagnosing certain diseases and also identifying certain species of fish.

Gills: *These are used to extract oxygen from the water and they pass over the fine gill filaments. Some fish use the gills to sift food from the water and substrate.*

Barbel: *A sensory organ that allows the fish to feel and 'smell' for food that may be hidden in the substrate.*

The Inner ear: *Used by the fish to sense their surroundings by interpreting vibrations and information received from the lateral line.*

Lateral line: *A sensory organ consisting of a linked row of sensory pits. These detect vibration and may be able to detect minute changes in electromagnetic fields.*

bunch of fish to spawn, well worth the effort. Once the fry hatch, feed them on infusorians and newly hatched brineshrimp.

CULTURING INFUSORIANS

Infusorians are tiny single-celled organisms which are found in almost any body of water. Their tiny size makes them an ideal first food for fry, and the ease of culture makes them accessible to anyone. It is best to carry out infusorian culture using three large glass jars. One should be started every three to four days to ensure a steady supply. Three-quarters fill them with cooled, boiled water and add three or four lettuce leaves that have been squeezed or bruised. Place the jar, with the lid off, where it is moderately well-lit and slightly warm. Over the next few days, the water will start to turn cloudy and begin to smell slightly. As the water clears and the smell goes, the infusorians begin to appear. They are not easily visible to the naked eye but will be found by the fry. When the water is completely clear, syphon a little of it into the fry tank. It is best to use airline for this job so that overfeeding is avoided. As the jar runs low, start the new culture in the same way.

HEALTH CARE

No matter how careful you are, your fish will suffer some kind of ailment at some time in their life. As with you and I, fish which are stressed are more likely to suffer illness than those that have a stress-free existence. Stress can be caused by anything from bullying to poor water quality. Either situation will reduce the fish's natural immune system until the fish falls ill. Fish diseases fall into three main categories – bacterial, fungal and parasitic. To fully understand the diseases we will look at each one separately.

BACTERIAL INFECTIONS

There are several strains of bacteria likely to affect your fish, and most of them are permanently present in the aquarium. Most often, bacteria infect the fish as a result of damage or poor water conditions. There is a whole batch of anti-bacterial treatments available. If caught early, most bacterial infections can be treated and easily cured. Internal bacterial infections are a little harder to detect, as their external symptoms do not show through until the disease is well established. As your fishkeeping knowledge grows, your understanding of your fish will increase and you will be able to notice when your fish are unwell. Once a disease is spotted, prompt treatment will bring about a rapid recovery. The following list includes some of the common bacterial infections.

FINROT

This is evident by the ragged appearance of your fish's fins. It should not be confused with split and damaged fins which can result in fish that are being bullied. Finrot on fish is shown by the tattered white fins with a red edge to the undamaged portion of the fin. It can result from bad handling, bullying or poor water conditions.

Treatment: Immediately carry out a partial water change with water treated with a de-chlorinator. Once the tank has been re-filled, remove any carbon filter media and dose with a proprietary anti-bacterial remedy. Depending on the remedy, the symptoms should clear up within a week or so, with the damaged fins fully repaired within a month or so.

ULCER DISEASE

These nasty open wounds or sores can be found all over the body of the fish but especially along the flank. These are common on goldfish but can affect all manner of fish. Other symptoms include loss of appetite, a build-up of body fluid and reddening at the base of the fins. Some secondary infections include fungal growths but it is not uncommon for the fish to die of this illness without any external symptoms showing. The bacteria responsible for this disease are always present in the aquarium but only take effect if the fish is run-down with poor water quality.

Treatment: Successful treatment is quite difficult as the disease is so virulent. Antibiotics are the best option but these can only be administered by a vet. If your fish is large and expensive, the cost involved will be worth it. Otherwise, buy one of the proprietary treatments available. If the fish shows no sign of improvement, it is more humane to painlessly destroy the fish. (See notes at the end of the section).

DISCOLOURED PATCHES

From time to time, your fish may develop discoloured patches on their skin. These are often slight bacterial infections which can easily be treated. If left, they may develop into something more serious. This type of bacterial infection is often the result of poor water quality and can only be prevented from recurring by improving water quality with regular partial water changes. The bacteria responsible for these types of infection are always present in the aquarium so if a fish is damaged by netting, or through bullying by another fish, it may contract a bacterial infection. If a fish is healthy and the tank has good water quality, an infection will not take a hold.

Treatment: Without good water quality, treatment is useless. A water change of around 15 per cent should be the first course of action followed by the correct dose of a broad range bactericide.

This will stop and improve the bacterial infection within a week. For the best results, rapid response is the key. Stop the infection as soon as it starts.

DROPSY AND POPEYE

This ailment is very easy to identify. The fish will take on a markedly bloated appearance with its scales standing out at right angles to the body. This gives the overall appearance of a pine cone. The fins will also redden at their bases and the faeces usually become long and pale. Popeye, as the name suggests, is where the fish's eyes swell and protrude from the side of the head. If left they may burst, the fish will survive and the wound will heal but the fish may be blind. This is usually the result of a number of factors but the root of the problem is often poor water quality. Other reasons for dropsy can be linked to an internal bacterial infection causing swelling of the internal organs.

Treatment: If only one or two individuals are affected, they should be removed and isolated so more intensive treatment can be administered. Daily partial water changes are thoroughly recommended and the addition of aquarium salt at a rate of 10 grams per litre will all help. Should the problem be an internal bacterial infection, there are remedies for this. While treating for these symptoms, feed your fish frozen food to the exclusion of everything else. This may prevent the intestinal tract becoming blocked, which is a common occurrence with a fish suffering from dropsy. If the whole tank of fish is suffering from the same ailments, regular partial water changes are needed at the rate of 10-15 per cent a day until the fish start to improve. Large and expensive fish may have dropsy cured with a simple antibiotic injection from a vet.

SWIM BLADDER PROBLEMS

These are very common in goldfish, particularly fancy varieties. An affected fish is usually out of control of its buoyancy to such an extent that it floats upside down at the surface and is unable to reach the bottom of the tank. This obviously causes a great deal of stress to the fish, which may become infected with a secondary illness. The causes of this ailment can be numerous, from a blocked intestine to an internal bacterial infection. Many fancy varieties of goldfish suffer these problems due to a deformed swim bladder owing to their unnatural shape. ***Treatment:*** Remove any infected fish to a separate tank. Ideally the water should be a few degrees warmer than the water in the stock tank. The higher temperature increases the metabolic rate and so may speed a recovery. Feed the fish solely on frozen food such as bloodworm and daphnia. These help to shift any blockage in the gut which may be putting pressure on the swim bladder. If, after a few days, the problem has not improved, treat the fish to a dose of anti-internal bacteria treatment. An improvement should be seen within a few days. If nothing happens, it is better painlessly to destroy the fish as it will only continue to suffer more stress.

'Normal' bodied goldfish are less prone to swimbladder problems than fancy varieties.

FUNGAL INFECTIONS

These are usually secondary infections that exacerbate existing bacterial infections or physical damage. Fungal infections are usually identified by cotton wool-like growths on the body or finnage. These infections should not appear, as they infect areas that are already damaged and which should not have gone unnoticed or untreated so, in theory, fungal infections should be a rare occurrence in the aquarium.
Treatment: Rapid treatment of fungal infections is easy. Broad-range fungicides are available and an improvement is usually seen within a week. For less serious infections, salt baths can be used. Simply fill an ice cream tub with tank-water and aquarium salt at the rate of 10 grams per litre. Slowly introduce the fish to the salt bath over a period of twenty minutes or so. Leave the fish in the bath for between 30 to 40 minutes before slowly diluting the bath back to full freshwater taken from the aquarium. Usually one bath is enough, so long as correct water quality is maintained in the aquarium. If the fungus is persistent, a second bath may be called for.

PARASITES

These come in a number of forms, from tiny external parasites which cause slimy skin, to large internal tapeworms several feet in length. Many parasites can be found on the average fish, and in small numbers, do not cause much harm. It is only in the confines of an aquarium or pond that parasites can be a problem. If a fish is seen to be carrying an external parasite, isolate it immediately as this will help to prevent the spread of the organism from fish to fish. The parasite, such as whitespot, is one of the most common diseases that fish-keepers are likely to encounter. Often a slight reduction in the fish's immune system is all that is required to encourage an attack of whitespot. There are a number of parasites which the fishkeeper may encounter, both external and internal. The following are a few of the most common.

WHITESPOT

These tiny white cysts measure around 1mm across and attach themselves all over the surface of the fish, usually starting on the fins. Fish with heavy infestations look as though they have been sprinkled with sugar grains and

Most fish harbour parasites on their bodies

will often be seen scratching against rocks or the gravel. Whitespot can be contracted by all fish, particularly those that have been subjected to stress, whether from poor water quality, temperature shock or bullying.

Treatment: The profusion of shop-bought remedies available for the treatment of whitespot is some indication as to the regular occurrence of this disease. Fully-developed whitespot parasites are impervious to treatment, so any remedy has to be administered over a period of 10 to 14 days. This encompasses the complete life-cycle of the whitespot parasite, from encysted larvae to an adult attached to the fish. Ensure that the remedy is dosed for the specified period, as missing a day may prolong the infestation.

FISH LICE

These disc-like parasitic crustaceans, measuring around 10mm across, are often brought into your tank with new fish. At the front of the lice is a pair of disc-like suckers which it uses to attach itself to the host fish so that it can feed on the fish's blood. When the lice moves on, the wound left by its probing mouth parts may become infected by bacterial or fungal spores. These are most common in ponds but do occur in tanks if introduced on new fish.

Treatment: Remedies are few and far between. The best treatment for fish lice infestations are particularly noxious chemicals that are used on commercial fish farms. Without specialist training and proper equipment, use of the chemicals is not possible. There are anti-parasite remedies which are available from shops and, though these may take a little longer to work, they still give an effective cure.

LEECHES

Leeches can be up to 2 ins long and are identified by their jointed, worm like bodies and disc-like suckers on each end of their body. When they move from fish to fish, they usually leave behind a reddened patch which may become infected by bacteria or fungus. They can be introduced to the tank or pond on new plants or fish; a careful look at the plant before putting it in the tank or pond can save you trouble in the long run. Although the leech itself does not do a great deal of damage, the wounds they inflict and the microbial disease they transmit can be deadly to your fish.

Treatment: Whatever treatment you choose to use, the eggs of the leech are not affected. Consequently more than one treatment is necessary. Most aquatic shops will stock some form of anti-parasite which will be effective for the treatment of leeches. The course of medication usually takes a week or two.

ANCHOR WORM

These rather nasty little creatures appear on the flanks of fish and resemble a large brown splinter. The head of the worm embeds itself into the muscle of the fish and can penetrate deep enough to damage internal organs. At the point of attachment an ulcer usually develops, where secondary infection takes a hold. Heavy infestation can cause dramatic weight loss and even death.
Treatment: These are fairly difficult to eradicate. The best procedure is manual removal with a fine-nosed pair of forceps or tweezers. Carefully net the fish and restrain it in a soft damp cloth. Once the adult worm has been removed, dab the damaged area on the fish with a broad-range bactericide and fungicide. The tank will need dosing with an anti-parasite remedy combined with careful observation over the next few weeks to make sure a second infestation does not occur.

INTESTINAL WORMS

These consist of three main types of worms which infest the gut. Thankfully they are not common unless the fish are wild-caught or newly imported. Symptoms include a grossly distended body or, conversely, a very thin and emaciated body. Sometimes the worms can be seen protruding from the fish's anus.
Treatment: For successful treatment of intestinal worms, a specific antihelminth treatment is required. These can only be acquired from a vet. No other remedy is available.